Applying the Concept of Wilderness Character to National Forest Planning, Monitoring, and Management

Landres, Peter; Hennessy, Mary Beth; Schlenker, Kimberly; Cole, David N.; Boutcher, Steve. 2008. **Applying the concept of wilderness character to national forest planning, monitoring, and management**. Gen. Tech. Rep. RMRS-GTR-217WWW. Fort Collins, CO: U.S. Department of Agriculture, Forest Service, Rocky Mountain Research Station. 45 p.

Abstract

The U.S. Forest Service is responsible for managing over 35 million acres of designated wilderness, about 18 percent of all the land managed by the agency. Nearly all (90 percent) of the National Forests and Grasslands administer designated wilderness. Although the central mandate from the 1964 Wilderness Act is that the administering agencies preserve the wilderness character in these designated areas, the concept of wilderness character has largely been absent in Forest Service efforts to manage wilderness. The purpose of this document is to help National Forest planners, wilderness staff, and project leaders apply in a practical way the concept of wilderness character to forest and project planning, the National Environmental Policy Act process, on-the-ground wilderness management, and wilderness character trend monitoring that is relevant to an individual wilderness.

Keywords: Wilderness Act, wilderness, wilderness character, wilderness stewardship, planning, monitoring

Acknowledgments

The authors greatly appreciate the help of Tom Carlson who wrote the Fire Resource Advisor section, and the review comments from Kevin Cannon, Tim Eling, Linda Merigliano, and Dave Parsons that helped clarify the ideas and writing of this paper. The authors also thank Kristi Coughlon for her sure and capable editing.

Rocky Mountain Research Station
Publishing Services

Telephone	(970) 498-1392
FAX	(970) 498-1122
E-mail	rschneider@fs.fed.us
Web site	http://www.fs.fed.us/rm/publications
Mailing Address	Publications Distribution
	Rocky Mountain Research Station
	240 West Prospect Road
	Fort Collins, CO 80526

The Authors

Peter Landres is an Ecologist at the Aldo Leopold Wilderness Research Institute, Rocky Mountain Research Station, U.S. Forest Service, Missoula, Montana. He holds a Ph.D. in ecology and biology from Utah State University and a B.S. in natural science from Lewis and Clark College.

Mary Beth Hennessy is the Forest Resource Staff Officer for the Inyo National Forest, U.S. Forest Service, Bishop, California. She holds a M.S. in environmental studies from the University of Montana and a B.A. in history from the University of California, Santa Barbara.

Kimberly Schlenker is the Wilderness and Recreation Program Manager for the Gallatin National Forest, U.S. Forest Service, Bozeman, Montana. She holds a B.S. in natural resource management from the University of Montana.

David Cole is a Geographer at the Aldo Leopold Wilderness Research Institute, Rocky Mountain Research Station, U.S. Forest Service, Missoula, Montana. He holds a Ph.D. in geography from the University of Oregon and an A.B.S. in geography from the University of California, Berkeley.

Steve Boutcher is the national Information Manager for the Wilderness & Wild and Scenic Rivers Staff, Washington Office, U.S. Forest Service and is based out of Burlington, Vermont. He holds a B.S. in forestry, with a Coordinate Degree in environmental studies from the University of Vermont.

Author's Note: This publication was developed by a technical working group and solely represents the views of its authors. It does not represent and should not be construed to represent any agency determination or policy.

Contents

Introduction... 1
  How to Use This Document 2

Wilderness Character 2
  The Importance of Wilderness Character......................... 2
  Application to Wilderness Stewardship 3
    Untrammeled quality 3
    Natural quality 4
    Undeveloped quality... 7
    Solitude or primitive and unconfined quality 9

Applying Wilderness Character to the NEPA Process............... 11
  Plan to Project: "Left Side of the Triangle" Analysis........... 12
    Desired condition/Existing condition 12
    Need for change 12
    Possible activities 14
  Scoping: "Right Side of the Triangle" Analysis.................. 14
    Proposed action and "Purpose and Need" 15
    Issue identification 16
    Alternatives.. 17
    Effects... 17
    Cumulative effects 19
  Decision Document 20

Monitoring Trend in Wilderness Character at the Local Level 21
  Process to Monitor Trend in Wilderness Character 21
    Step 1 — Review the list of indicators........................ 21
    Step 2 — Identify measures for each indicator 21
    Step 3 — Assess trend for each measure 22
    Step 4 — Assess trend for each indicator..................... 22
    Step 5 — Assess trend for each quality of wilderness character 24
    Step 6 — Assess overall trend in wilderness character 24
  Monitoring and Information Management......................... 24

Applying Wilderness Character to Managing Wilderness........... 25
  Minimum Requirements Decision Guide.......................... 25
  Fire Resource Advisor 26
  Information Needs Assessment 27
  Work Planning 27

References... 28

Appendix A—Indicators and Example Measures for the Four Qualities.......... 30

Appendix B—Hypothetical Decision Memo Scenario 32

Appendix C—Summarizing Effects Using Wilderness Character Indicators....... 43

Applying the Concept of Wilderness Character to National Forest Planning, Monitoring, and Management

Peter Landres, Mary Beth Hennessy, Kimberly Schlenker,
David N. Cole, and Steve Boutcher

Introduction

The U.S. Forest Service is responsible for managing over 35 million acres of designated wilderness, about 18 percent of all the land managed by the agency. Nearly all (90 percent) of the National Forests and Grasslands administer designated wilderness. Although the central mandate from the 1964 Wilderness Act is that the administering agencies preserve the wilderness character in these designated areas, the concept of wilderness character has largely been absent in Forest Service efforts to manage wilderness. The purpose of this document is to help National Forest planners, wilderness staff, and project leaders apply in a practical way the concept of wilderness character to forest and project planning, on-the-ground wilderness management, and wilderness character trend monitoring that is relevant to an individual wilderness.

The ideas in this document are based on the Forest Service's national framework to monitor selected conditions related to wilderness character (Landres and others 2005), the Technical Guide for monitoring these conditions (Landres and others, in press), and the recently released interagency strategy for monitoring trends in wilderness character (Landres and others 2008).

Applying the concept of wilderness character to National Forest planning, management, and monitoring should be directly useful to improving on-the-ground wilderness stewardship in several ways:

- encouraging a comprehensive, holistic look at important wilderness attributes that are directly linked to the concept of wilderness character;
- building support for preserving wilderness character directly into forest and project planning;
- monitoring trends in wilderness character that are relevant to the individual wilderness and the forest;
- communicating stewardship needs and priorities related to wilderness character within the agency and with the public; and

- aiding in the fulfillment of Element 8 ("This wilderness has adequate direction in the Forest Plan to prevent degradation of the wilderness resource") and Element 9 ("The priority information needs for this wilderness have been addressed through field data collection, storage, and analysis") of the U.S. Forest Service's 10-Year Wilderness Stewardship Challenge (Wilderness Advisory Group 2008).

How to Use This Document

This document is intended to demonstrate the direct and practical application of wilderness character concepts to various wilderness stewardship activities. Examples are provided for many activities, but not all. The reader is encouraged to be creative and apply these concepts in many ways to forest level wilderness stewardship. Managing to preserve wilderness character may be a new approach to some, and much can be gained from simply trying to incorporate these concepts at the local level.

Wilderness Character

This section briefly describes why the concept of wilderness character is important to forest level planning and management, and provides detailed information about the four qualities of wilderness character. Those wanting to understand the "how" but not the "why" are encouraged to at least skim the following sections and review the list of indicators and sample measures included under each of the four qualities of wilderness character in this section. Skimming these sections will provide an overview of foundational wilderness character concepts before proceeding to the "Applying Wilderness Character to the NEPA Process" and subsequent sections. A complete listing of all the indicators and sample measures is given in *Appendix A*.

The Importance of Wilderness Character

The Wilderness Act of 1964, Use of Wilderness Areas Section 4(b), describes the primary direction for wilderness stewardship as "each agency administering any area designated as wilderness shall be responsible for preserving the wilderness character of the area" (McCloskey 1999; Rohlf and Honnold 1988; United States Congress 1983). Agency wilderness policy, Forest Service Manual 2320.2 (4), directs the agency to "protect and perpetuate wilderness character" from the time of wilderness designation. The central importance of preserving wilderness character is underscored by several recent District Court and Court of Appeals cases showing that the courts are increasingly holding the agency accountable for fulfilling this legal and policy mandate to preserve wilderness character.

In addition to law and policy, focusing on wilderness character links on-the-ground wilderness conditions to the mandates of the Wilderness Act and agency policy, helping managers to:

- understand how stewardship decisions influence trends in wilderness character;
- improve agency defensibility in legal questions regarding preservation of wilderness character;
- establish priorities for stewardship actions that show the most promise to improve the trend in wilderness character; and
- provide a powerful communication tool to easily convey whether or not the agency is preserving wilderness character.

Application to Wilderness Stewardship

The Wilderness Act of 1964 doesn't define wilderness character and there is no discussion about its meaning in the legislative history of this act (Scott 2002). The Forest Service national framework for monitoring wilderness character (Landres and others 2005) concluded that wilderness character is ideally described as the unique combination of a) natural environments that are relatively free from modern human manipulation and impacts, b) opportunities for personal experiences in environments that are relatively free from the encumbrances and signs of modern society, and c) symbolic meanings of humility, restraint, and interdependence in how individuals and society view their relationship to nature.

Using the Definition of Wilderness, Section 2(c) from the Wilderness Act of 1964, the Forest Service national framework (Landres and others 2005) and Technical Guide (Landres and others, in press) identify four qualities of wilderness that make the idealized description of wilderness character relevant, tangible, and practical to forest planning, management, and monitoring. These four qualities can be applied to the stewardship of all National Forest System wildernesses—regardless of size, location, or other unique place-specific attributes—because they are based on the legal definition of wilderness and every wilderness law includes specific language that ties it to this definition (Hendee and Dawson 2002, Landres 2003). In addition to the four qualities of wilderness character that are discussed in detail below, there are also important intangible aspects (Keeling 2007, Schroeder 2007) such as the feelings of inspiration and spiritual connection.

Untrammeled quality—The Wilderness Act, Section 2(c) states that wilderness is "hereby recognized as an area where the earth and its community of life are untrammeled by man." The word "untrammeled" is rarely used in ordinary conversation, but Howard Zahniser, the primary author of the Wilderness Act, used untrammeled as a key word in the definition of wilderness.

USDA Forest Service Gen. Tech. Rep. RMRS-GTR-217WWW. 2008

3

Since passage of the Act, the word untrammeled and its meaning for wilderness stewardship have been discussed at length (for example, Aplet 1999, Scott 2002). Untrammeled means "allowed to run free" (American Heritage Dictionary 2006). Synonyms for untrammeled include unrestrained, unmanipulated, unrestricted, unhindered, unimpeded, unencumbered, self-willed, and wild.

Zahniser (1963) noted that the inspiration for wilderness preservation "is to use 'skill, judgment, and ecologic sensitivity' for the protection of some areas within which natural forces may operate without man's management and manipulation." Wilderness is very different than other lands in that legislation dictates not only the goals of stewardship, but how management is to be approached—with humility and with an eye toward not interfering with nature and not manipulating the land and its community of life.

Actions that intentionally manipulate or control ecological systems inside wilderness degrade the untrammeled quality of wilderness character, even though they may be taken to restore natural conditions. For example, wilderness is manipulated and the untrammeled quality of wilderness character is diminished when naturally ignited fires are suppressed inside wilderness, dams are built that impede natural water flow, animals or plants are removed, or landscapes are restored by removing trees or introducing trees that are genetically resistant to pathogens. This concept of trammeling applies to all manipulation since the time of wilderness designation. It does not apply to manipulations that occurred prior to wilderness designation, such as the use of fire by native people to promote game habitat, because the mandates of the Wilderness Act don't apply prior to designation.

Unlike the management of any other land in the Nation, wilderness legislation directs the managing agency to scrutinize its actions and minimize control or interference with plants, animals, soils, water-bodies, and natural processes. Prominence of "untrammeled" in the Wilderness Act distinguishes the untrammeled quality from the natural quality, although the two are clearly linked. In essence, the untrammeled quality monitors *actions* that intentionally manipulate or control ecological systems, whereas the natural quality monitors the intentional and unintentional *effects* from actions taken inside wilderness as well as from external forces on these systems. Separating actions from effects offers clearer understanding of trends in actions compared to trends in effects, permitting more effective analysis and use of the information to improve wilderness stewardship.

Indicators and sample measures under the untrammeled quality are provided in table 1.

Natural quality—One of the major themes running throughout the 1964 Wilderness Act is that wilderness should be free from the effects of "an increasing population, accompanied by expanding settlement and growing mechanization" and that the "earth and its community of life...is protected and managed so as to preserve its natural conditions" (Sections 2(a) and 2(c), respectively). Historically, wilderness is strongly associated with protecting and preserving ecological systems from the impacts of modern people (Sutter 2004).

Table 1—Indicators and sample measures for the untrammeled quality.

Quality	Indicator	Example measures
Untrammeled — Wilderness is essentially unhindered and free from modern human control or manipulation	Actions authorized by the Federal land manager that manipulate the biophysical environment	Number of actions to manage plants, animals, pathogens, soil, water, or fire
		Percent of natural fire starts that received a suppression response
		Number of lakes and other water bodies stocked with fish
	Actions not authorized by the Federal land manager that manipulate the biophysical environment	Number of unauthorized actions by other Federal or State agencies, citizen groups, or individuals that manipulate plants, animals, pathogens, soil, water, or fire

In today's terms, this means that the indigenous species composition and the structures and functions of the ecological systems in wilderness are protected and allowed to be on their own, without the planned intervention or the unintended effects of modern civilization. Only through such protection may wilderness truly serve as "a laboratory for the study of land-health" (Leopold 1949) and as an ecological baseline for understanding the effects of modern civilization on natural systems.

Ecological systems inside wilderness are directly affected by things that happen inside, as well as outside the wilderness, and by actions taken by agencies or citizens inside wilderness. For example, non-indigenous fish are intentionally introduced for recreational fishing, yet have far-reaching unanticipated negative effects on native biological diversity and nutrient cycling in wilderness lakes (Knapp and others 2001). Livestock grazing may be allowed in wilderness but may contribute to soil disturbance and the spread of non-indigenous plants (Belsky and Blumenthal 1997). Biological control agents may be used to eradicate invasive non-indigenous plants but may have unintended effects on indigenous plants (Louda and Stiling 2004). Dams outside wilderness alter hydrological flow regimes, adversely affecting the riparian plant communities within wilderness (Cowell and Dyer 2002). Air pollutants from sources outside wilderness disperse long distances, affecting wilderness vegetation, soils, and aquatic systems (Schreiber and Newman 1987). Every wilderness shows the impacts of becoming increasingly isolated within a "sea" of modern development (Landres and others 1998).

All ecological systems change over time and vary from one place to another, and neither law nor policy are intended to maintain static or unchanging natural conditions in wilderness. Separating anthropogenic change from natural change implies that there is sufficient understanding about how ecological systems naturally vary over time and across a landscape to separate human-caused from natural change. In practice, this understanding is lacking in most areas.

USDA Forest Service Gen. Tech. Rep. RMRS-GTR-217WWW. 2008

5

Indicators and sample measures under the natural quality are provided in table 2.

Table 2—Indicators and sample measures for the natural quality.

Quality	Indicator	Example measures
Natural — Wilderness ecological systems are substantially free from the effects of modern civilization	Plant and animal species and communities	Abundance, distribution, or number of indigenous species that are listed as threatened and endangered, sensitive, or of concern
		Number of extirpated indigenous species
		Number of non-indigenous species
		Abundance, distribution, or number of invasive non-indigenous species
		Number of acres of authorized active grazing allotments and number of animal unit months (AUMs) of actual use inside wilderness
		Change in demography or composition of communities
	Physical resources	Visibility based on average deciview and sum of anthropogenic fine nitrate and sulfate
		Ozone air pollution based on concentration of N100 episodic and W126 chronic ozone exposure affecting sensitive plants
		Acid deposition based on concentration of sulfur and nitrogen in wet deposition
		Extent and magnitude of change in water quality
		Extent and magnitude of human-caused stream bank erosion
		Extent and magnitude of disturbance or loss of soil or soil crusts
	Biophysical processes	Departure from natural fire regimes averaged over the wilderness
		Extent and magnitude of global climate change based on change in timing of greening from MODIS satellite imagery, glacial retreat from photopoints, change in temperature and precipitation patterns from RAWS data, change in snow depth from SNOTEL data, coastal erosion or accretion from photopoints, or change in the distribution of select plant communities (for example, treeline) from photopoints

Undeveloped quality—Wilderness is defined in Section 2(c) of the 1964 Wilderness Act as "an area of undeveloped Federal land retaining its primeval character and influence, without permanent improvements or human habitation," with "the imprint of man's work substantially unnoticeable." The basic idea that wilderness is undeveloped runs through every definition of wilderness. For example, Aldo Leopold (1921) envisioned wilderness as "a continuous stretch of country preserved in its natural state, open to lawful hunting and fishing, devoid of roads, artificial trails, cottages, or other works of man." Hubert Humphrey (1957), an original sponsor of the Wilderness Act, clarified his definition of wilderness as "the native condition of the area, undeveloped... untouched by the hand of man or his mechanical products."

The Wilderness Act identifies "expanding settlement and growing mechanization" as forces causing wild country to become occupied and modified, and further clarifies in Section 4(c) that "there shall be no temporary road, no use of motor vehicles, motorized equipment or motorboats, no landing of aircraft, no other form of mechanical transport, and no structure or installation." An early Forest Service review of wilderness policy (USDA Forest Service 1972) noted that buildings or structures are usually installed for only one purpose— to facilitate human activity. The building or structure not only occupies the land, but makes it easier for people to impose their will on the environment, thereby modifying it. This policy review also found that motorized equipment and mechanical transport similarly make it easier for people to occupy and modify the land. Zahniser (1956) articulated this idea when he argued the need for "areas of the earth within which we stand without our mechanisms that make us immediate masters over our environment." While the use of motorized equipment or mechanical transport affects the opportunity for visitors to experience natural quiet and primitive recreation, these uses are included under this undeveloped quality due to their close association with people's ability to develop, occupy, and modify wilderness.

No wilderness has escaped at least some modern human occupation and modification. Many developments were "grandfathered" into the wilderness by special provisions in the enabling legislation, including buildings, roads, dams, powerline and water pipe corridors, and mines. While the continuing presence of these developments may be legal uses of wilderness, the resulting facilities, structures, and authorizations for motorized use and mechanical transport can have far-reaching effects on wilderness character (Hendee and Dawson 2002). The variety of special provisions that are unique to each wilderness underscore the importance of not comparing one wilderness to another.

Many developments degrade both the undeveloped quality and the solitude or primitive and unconfined recreation quality. Following the Interagency Strategy to Monitor Trends in Wilderness Character, all non-recreational developments (such as administrative sites, dams, stock fencing, fixed instrumentation sites, or trails and roads used to access inholdings) are included in the undeveloped quality. All recreation-focused developments (such as trails, campsites, shelters, or toilets) are included in the solitude or primitive and unconfined recreation quality.

Heritage or cultural resources within a wilderness may be an important part of wilderness character. Including heritage resources as part of wilderness character is controversial (see the Interagency Strategy for Monitoring Trends in Wilderness Character for a detailed discussion). We include them in this document to alert readers to the possibility of their inclusion in wilderness character. We include them in the undeveloped quality because they primarily represent human relationships with the land prior to modern wilderness designation, directly supporting the basis for the undeveloped quality as explained above. Wilderness and heritage resources staffs have, at times, disagreed about what are considered significant heritage resources and how to manage them in wilderness. These are local decisions, and we stress that local staffs, using both the Wilderness Act and cultural resource protection laws, should work together to develop a common understanding for which heritage resources will be considered as part of preserving wilderness character.

Indicators and sample measures in the undeveloped quality are provided in table 3.

Table 3—Indicators and sample measures for the undeveloped quality.

Quality	Indicator	Example measures
Undeveloped — Wilderness retains its primeval character and influence, and is essentially without permanent improvement or modern human occupation	Non-recreational structures, installations, and developments	Index of authorized physical development
		Index of unauthorized (user-created) physical development
	Inholdings	Area and existing or potential impact of inholdings
	Use of motor vehicles, motorized equipment, or mechanical transport	Type and amount of administrative and non-emergency use of motor vehicles, motorized equipment, or mechanical transport
		Type and amount of emergency use of motor vehicles, motorized equipment, or mechanical transport
		Type and amount of motor vehicle, motorized equipment, or mechanical transport use not authorized by the Federal land manager
	Loss of statutorily protected cultural resources	Number and severity of disturbances to cultural resources

Solitude or primitive and unconfined quality—The Wilderness Act states in Section 2(c) that wilderness has "outstanding opportunities for solitude or a primitive and unconfined type of recreation." The intended meaning of this wording by the framers of the Wilderness Act isn't recorded in the legislative history of the Act and it has caused much discussion and debate among wilderness managers and scholars (Hendee and Dawson 2002). However, early wilderness writings of Aldo Leopold, Robert Marshall, Howard Zahniser, and others paint a rich picture of the type of experience envisioned in wilderness environments (see Landres and others 2005 for examples). These writings strongly enforce the vital role of solitude in places that are primitive and unconfined as central to the idea of wilderness.

The meaning of solitude has been at the center of considerable debate among researchers and the public. Such meanings range from a lack of seeing other people to privacy, freedom from societal constraints and obligations, and freedom from management regulations (Hall 2001). Given the content of early wilderness writings, it is likely that solitude was viewed holistically, encompassing attributes such as separation from people and civilization, inspiration (an awakening of the senses, connection with the beauty of nature and the larger community of life), and a sense of timelessness (allowing one to let go of day-to-day obligations, go at one's own pace, and spend time reflecting).

Primitive and unconfined recreation has also been the subject of much debate. Primitive recreation has largely been interpreted as travel by non-motorized and non-mechanical means (for example by horse, foot, or canoe) that reinforces the connection to our ancestors and American heritage. However, primitive recreation also encompasses reliance on personal skills to travel and camp in an area, rather than reliance on facilities or outside help (Roggenbuck 2004). Unconfined encompasses attributes such as self-discovery, exploration, and freedom from societal or managerial controls (Hendee and Dawson 2002, Lucas 1983). Primitive and unconfined environments together provide ideal opportunities for the physical and mental challenges associated with adventure, real consequences for mistakes, and personal growth that result from facing and overcoming obstacles (Borrie 2000, Dustin and McAvoy 2000).

In certain situations, managers may need to make a difficult decision about the need for resource protection while providing outstanding opportunities for primitive and unconfined recreation. For example, administrative sites or a minimal system of trails may be considered essential to manage the effects of recreation use while still allowing people to use and enjoy wilderness. However, since structures and system trails may strongly influence people's opportunity for primitive and unconfined wilderness recreation, the agencies need to show restraint in fulfilling their administrative responsibilities so that the primitive and unconfined quality of wilderness does not slowly erode over time.

Many different factors contribute in known and unknown ways to the experience of solitude or primitive and unconfined recreation (Borrie and Birzell 2001; Hendee and Dawson 2002; Manning and Lime 2000). For example, experiences may be influenced by factors largely beyond the control and influence of managers. Such factors include the attributes of the physical landscape, presence of certain animals (for example, mosquitoes and grizzly bears), local weather, intra- and inter-group dynamics, and skills and knowledge an individual brings to the experience. In contrast, managers may exert some control over use levels, types and patterns of use, level of development (both inside and adjacent to wilderness), amount and type of information available about the wilderness, and types of regulations imposed, all of which influence the opportunity to experience solitude or a primitive and unconfined type of recreation (Cole and others 1987; Lucas 1973; McDonald and others 1989; Watson 1995).

Indicators and sample measures under the solitude or primitive and unconfined recreation quality are provided in table 4.

Table 4—Indicators and sample measures for the solitude or primitive and unconfined quality.

Quality	Indicator	Example measures
Solitude or Primitive and Unconfined Recreation — Wilderness provides outstanding opportunities for solitude or primitive and unconfined recreation	Remoteness from sights and sounds of people inside the wilderness	Amount of visitor use
		Number of trail contacts
		Number and condition of campsites
		Area of wilderness affected by access or travel routes that are inside the wilderness
	Remoteness from occupied and modified areas outside the wilderness	Area of wilderness affected by access or travel routes that are adjacent to the wilderness
		Night sky visibility averaged over the wilderness
		Extent and magnitude of intrusions on the natural soundscape
	Facilities that decrease self-reliant recreation	Type and number of agency-provided recreation facilities
		Type and number of user-created recreation facilities
	Trail development level	Number of trail miles in developed condition classes
	Management restrictions on visitor behavior	Type and extent of management restrictions

Applying Wilderness Character to the NEPA Process

This section provides suggestions for how to incorporate wilderness character concepts into the National Environmental Policy Act (NEPA) process. While generally following what is known as the NEPA Triangle (fig. 1), this section is not a primer on NEPA and assumes a functional understanding of the NEPA process. Although wilderness character concepts can be applied to many aspects of the NEPA process, its greatest application may be in analyzing effects at the project level and describing desired conditions for programmatic direction in Forest or Wilderness Plans. This section demonstrates how the four qualities of wilderness character could be incorporated into NEPA at both programmatic and project levels. Ten hypothetical examples, in boxes, show how these qualities may be incorporated into each step of NEPA.

Management decisions require articulating the tradeoffs and expected results for projects that affect the preservation of wilderness character. For example, the decision not to build a footbridge across a stream may preserve certain qualities of wilderness character, while building the footbridge may diminish certain qualities of wilderness character. These distinctions are important to describe in the environmental compliance process, and the wilderness character concepts discussed here provide methodology and language to articulate how wilderness character is affected.

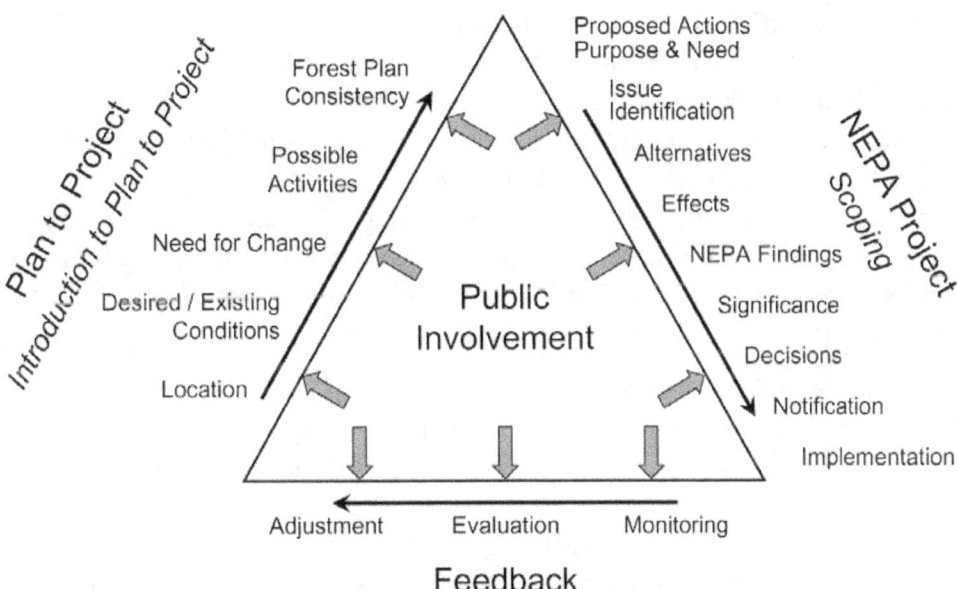

Figure 1—NEPA Triangle.

Wilderness designation sets the standard for managing an area to preserve all aspects of wilderness character. Given the tradeoffs that a decisionmaker may face, the NEPA process should not be used to justify preserving one quality of wilderness character at the expense of another. Instead, the NEPA process serves to illuminate impacts that will likely occur given certain actions, and the approach outlined here suggests a process for identifying how such actions may affect each of the four qualities of wilderness character. Ideally, the agency would prohibit any authorized action that degrades wilderness character and the NEPA process would guide the agency in making decisions that uphold this ideal.

Plan to Project: "Left Side of the Triangle" Analysis

The left side of the triangle occurs prior to formally entering the NEPA process. This is the project development process, which ends when a proposed action has been developed and is "ripe" for action. During this time, wilderness character considerations can be framed and subsequently developed and carried through the NEPA process, such as:

- What qualities of wilderness character are going to be affected by the project?
- Can the project be adjusted to mitigate potential impacts on wilderness character during the project development phase?

Desired condition/Existing condition—The first step is to compile all the current direction that exists for the planning area by reviewing Forest and Wilderness Plan direction, including amendments, and applicable Forest Service Manual direction. While many standards and guidelines and desired condition statements may not have historically incorporated wilderness character language, one can easily frame the desired condition in terms of the four qualities and track them through analysis. Examples of desired condition statements for each of the four qualities are given in Box 1.

Arranging current Forest Plan direction as demonstrated in Box 1 may help fulfill Element 8 of the Chief's 10-Year Wilderness Stewardship Challenge (Wilderness Advisory Group 2008) by evaluating that direction and identifying needs for additional direction for that wilderness.

Need for change—A proposal or action should come out of a need for change. Need for change is identified through a comparison of existing and desired conditions. An example of "uncovering" this need for change is given in Box 2.

The preliminary work of identifying the difference between existing and desired conditions will be further developed into the Purpose and Need for the project. See *Appendix B – Hypothetical Decision Memo Scenario* for an example of modifying need for change language into the Purpose and Need Statement.

Box 1 — Hypothetical Examples of Current Direction for the Four Qualities of Wilderness Character

Untrammeled

All administrative actions using a prohibitive use will have a minimum requirements analysis that indicates that it is the minimum necessary for protecting wilderness (1988 Smokey NF Land Management Plan).

Natural

Manage boreal toad breeding habitat to prevent disturbance (1988 Smokey NF Land Management Plan).

Preserve, restore, or enhance special aquatic features, such as meadows, lakes, ponds, bogs, fens, and wetlands, to provide ecological conditions and processes needed to recover or enhance the viability of species that rely on these areas (2003 Smokey NF LRMP Amendment #4).

Undeveloped

Trails will be constructed and maintained in Wilderness with native materials (1988 Smokey NF Land Management Plan).

Solitude or Primitive and Unconfined Recreation

Provide a range of opportunities for use and solitude across the wilderness landscape. Allow for recreation use in popular destinations and ensure that in areas of concentrated use, that use does not expand or enlarge spatially (1988 Smokey NF Land Management Plan).

Box 2 — Hypothetical Example of an Interdisciplinary Team Uncovering the Need for Action by Comparing and Contrasting Existing and Desired Condition

Desired Condition

- Natural
  - Preserve, restore, or enhance special aquatic features, such as meadows, lakes, ponds, bogs, fens, and wetlands, to provide ecological conditions and processes needed to recover or enhance the viability of species that rely on these areas (2003 Smokey NF LRMP Amendment #4).

Existing Condition

- Natural
  - Meadow has a large headcut in the lower portion of the southeast corner and has the potential to migrate upstream and cause a change in the course of the stream.

Need for Change

- Erosion associated with the headcut needs to be reduced and/or eliminated to preserve natural quality of wilderness character.

Possible Actions

- Reroute trail that is causing a nickpoint in the stream at the headcut
- Place rock in headcut to arrest erosion
- Close trail to use
- Build a small bridge over stream crossing
- Monitor headcut to see if it migrates or changes over time

Possible activities—The Need for Change (in this case to preserve the natural quality of wilderness) allows an ID team to consider all the possible actions that could accomplish the identified need for change. Alternative ways to accomplish the project that maintain or enhance wilderness character should be explored as a precursor to fully developing the proposed action. In this stage, it is important to determine the overall appropriateness of the project in wilderness and preliminarily assess the potential effects to the four qualities of wilderness character. By going through this initial exercise, some projects may be screened out as inappropriate in wilderness or impacts to other qualities of wilderness character may be revealed. This step should narrow the scope of potential actions and lead to one proposed action. An example of this fine-tuning is given in Box 3.

At this point, the appropriate line officer will have the interdisciplinary team further develop the proposed action that is his or her preferred course of action. Once an action is proposed and scoping occurs, the project formally starts the NEPA process and moves to the right side of the NEPA Triangle.

Box 3 — Hypothetical Example of a Wilderness Project that was Fine-Tuned Prior to Developing the Final Proposed Action

Agency engineers determined that the West Fork of Quiet Waters Bridge, a substantial packstock bridge, was failing and proposed to replace it. Prior to initiating NEPA, an interdisciplinary team reviewed the project on site. The initial intent of the visit was to determine the "minimum tool" for replacing the bridge — flying in stringers, using on-site native materials, or packing in a glue-laminated bridge. The original project proposal was to replace the bridge, not consider other options for the crossing. However, after reviewing the site and agreeing that a ford was a reasonable and safe accommodation on this particular stream crossing, the original project proposal to replace the bridge was dropped. Instead, the old bridge was removed, a stock ford was hardened and improved, and a small foot log for pedestrians was installed to facilitate high water crossings. The net result was improving the undeveloped quality of wilderness by removal of an obtrusive structure while maintaining opportunities for primitive and unconfined recreation. Additionally, by scaling back the scope of the project, lengthier NEPA analysis was avoided and the decision was documented with a categorical exclusion.

Scoping: "Right Side of the Triangle" Analysis

The right side of the NEPA Triangle details the steps of the NEPA process, from the initial project scoping through implementation. This side depicts several different phases, most of which are applicable to both project and programmatic NEPA processes.

Proposed action and "Purpose and Need"—Carefully constructing the proposed action and Purpose and Need for the action narrows the scope of analysis. A good Purpose and Need directly addresses three crucial questions: What change is needed? Why here? Why now? The answers to these questions should be outcomes of the left side process where a need for change drove a specific action.

Structuring Purpose and Need for action around wilderness character qualities can provide a framework that can be applied to subsequent parts of the NEPA process. At the programmatic planning level, the "Why here?" and "Why now?" should capture the elements of direction that are needed and why. Always ask the question, "Under what qualities of wilderness character do these needs fall?" This packaging ties actions to legislative goals and can be an important connection that can be tracked and traced throughout the remaining NEPA process. Examples of this need for change language at the programmatic level are given in Box 4.

Box 4 — Hypothetical Examples of Programmatic NEPA Purpose and Need Statement

Purpose and Need Statement for the Natural Quality

One of the purposes of this plan is to ensure that the natural quality of wilderness is preserved. Ecological processes have been adversely affected by management actions in the past. Several examples include: trail alignment projects that route trails through fragile meadow environments and revegetation of areas with nonindigenous species prior to designation.

Purpose and Need Statement for the Solitude Quality

One of the purposes of this plan is to enhance opportunities for solitude. Current conditions show many areas where the level of use is impairing visitor's ability to find opportunities for solitude during much of the summer. The forest has received many comments from the public of crowding and concerns with use levels. There is a need to ensure that levels of use are consistent with the character quality of providing opportunities for solitude or a primitive and unconfined type of recreation.

In project level or site-specific NEPA, it will be important to point back to desired condition statements when framing the need for action. An example of such a statement for the natural quality is given in Box 5.

Box 5 — Hypothetical Example of a Project Level Purpose and Need Statement

Purpose and Need Statement for the Natural Quality

There is a need to treat a large headcut resulting from historic livestock grazing in Moo meadow. This active headcut, if not stabilized, could cause erosion upstream as well as continue downstream erosion, bank instability, and water quality concerns affecting the natural quality of wilderness character. The stream is important habitat for the threatened moo chub. The proposed headcut treatment will preserve the natural qualities of wilderness character.

If adequate programmatic direction does not exist, or if there is no desired condition statement that makes sense for the project, consider framing the Purpose and Need Statement in terms of how the project maintains or improves wilderness character. This may be challenging in the case of some proponent-driven projects that are permissible in wilderness by special provision (see Section 4(d) of the Act).

Up to this point, all processes described are applicable to project and programmatic NEPA, whether using a Categorical Exclusion (CE) or conducting an Environmental Analyses (EA) or Environmental Impact Statements (EIS). If the project fits into one of the categories of actions excluded from documentation, then the appropriate level of NEPA may be a Decision Memo with or without a case file (see FSH 1909.15 for the level of documentation needed for each category). *Appendix B – Decision Memo Scenario* demonstrates the application of wilderness character concepts in a decision memo when no further analysis is needed and when a finding on "no effect to an extraordinary circumstance" is used.

Issue identification—Issues are a point of debate, dispute, or disagreement regarding anticipated effects of the proposed action, and there are outcomes of public scoping. Issues will drive alternatives. It is possible that an issue that has been raised also ties to a wilderness character quality, though the public may not have explicitly stated so. When describing an issue, convey the disagreement or controversy and tie it to wilderness character. An example of such an issue is given in Box 6.

Box 6 — Hypothetical Example of an Issue in a Wilderness Project Level NEPA

During scoping, a local botanist provided feedback to the Agency that their proposal to designate and manage wilderness campsites in the Blue Sky Lake basin may conflict with objectives stated in the Forest Plan to protect plant habitat of sensitive or listed species. The botanist is concerned that several of the proposed designated sites are located in habitat very similar to nearby locations where the threatened Blue Sky lily grows. The Interdisciplinary Team (IDT) did not identify this as a concern prior to scoping. Once public comments were reviewed, the issues identified by the IDT were amended to include concerns about the potential effect of the proposed action to naturalness and the potential impact to a listed species because the project has the potential to affect habitat of the Blue Sky lily. Indicators that might be used to describe the effects of each alternative include the loss of suitable Blue Sky lily habitat or direct effects to plant populations and the resulting effect to naturally functioning wilderness ecosystems.

Issue

The natural quality of wilderness character may be affected by the designation of campsites at Blue Sky Lake. The habitat of Blue Sky lily, a sensitive species, may be disturbed if sites are designated in certain locations where known individuals of this plant population have been identified.

Indicators to be Used in Analysis of Effects

Acres of habitat and numbers of individuals directly effected.

Alternatives—Alternatives are developed to respond to issues identified during scoping. Alternatives must also meet the Purpose and Need. So if the Purpose and Need is constructed favoring certain qualities of wilderness character, and issues point out conflicts with other qualities of wilderness character, the alternative(s) will find an approach that addresses the conflict while meeting the Purpose and Need. To ensure that the appropriate number of alternatives is analyzed, determine if there really is a difference in effects or use mitigation measures to address issues. If simple mitigations could offset the anticipated effects to wilderness character, these actions could be incorporated into the proposal instead of by generating separate alternatives. Referring to the example in Box 6, up-front mitigation could state something like "no campsites will be designated in suitable Blue Sky lily habitat without sensitive plant surveys prior to designation." This tactic would negate the need to develop alternatives that avoid critical habitat. This simple mitigation would ensure that there are no unanticipated effects to naturalness from the project as they relate to this sensitive plant. In some cases, mitigation may need to be applied to all action alternatives to reduce the number of alternatives that need to be analyzed.

Effects—In this phase, the direct, indirect, and cumulative effects for each alternative are disclosed. Often, the affected environment section (current or existing condition) is combined with the effects analysis, but sometimes they are separate chapters. Existing conditions are used as a baseline and/or a comparison for effects. To determine the need for change, existing conditions are compared with desired conditions (see Box 2). One can use the existing conditions identified and articulated in this process as a starting point for the affected environment portion of the analysis. Using wilderness character concepts can make this task easier and clearer by framing the affected environment and effects analysis in terms of the four wilderness character qualities.

Planners recommend framing the effects analyses in time and space. One way to provide this framing is to explain the methodology used to analyze wilderness effects and describe the type of impact or effect and its context, intensity, and duration. An example of explaining this methodology is given in Box 7.

Using the methodology of Box 7, impacts or effects can be described with words that convey the intensity of the impact, the temporal and spatial scales of this impact, and a direct link to wilderness character principles that become a standard for analysis. Examples of programmatic and project statements that include these impact reducing standard descriptions are given in Box 8.

Appendix A provides a tabular listing of the indicators for the four wilderness character qualities. Summary direct and indirect effects for each alternative considered may be aggregated in this table for project or programmatic effects analysis to provide a comparison of effects between alternatives and a summary of overall effect to wilderness character.

Box 7 — Hypothetical Example Methodology Discussion From an Effects Analysis

The wilderness resource discussion in this chapter will evaluate the effects of management actions on wilderness character using the four qualities of wilderness character. For the purpose of this analysis, the following approach is used:

Type of effect

Impacts were evaluated in terms of whether they would be beneficial (enhance one or more of the qualities of wilderness character) or adversely affect one or more of the qualities of wilderness character.

Context

Local effects are those that occur at site-specific locations within the wilderness. Regional effects would be impacts to a wilderness character quality on adjacent lands such as the adjacent Loon State Wildlife Reserve. There are activities outside the wilderness boundary of the Blue Arch Wilderness that may contribute to cumulative effects.

Intensity

The intensity of the impact considers whether the effect to wilderness character is negligible, minor, moderate, or major. Negligible effects are considered not detectable to the visitor and therefore expected to have no discernible outcome. Minor effects are slightly detectable though not expected to have overbearing results on wilderness character. Moderate effects would be clearly detectable to the visitor and could have an appreciable effect on one or more aspects of wilderness character. Major effects would have highly noticeable influence on the visitors experience and could permanently alter more than one aspect of wilderness character.

Duration

The duration of the effect considers whether the impact would occur in a short- or long-term period. Short-term effect on solitude, for example, would be temporary in duration, such as an encounter while traveling or camping. Long-term effect would have lasting effects on the wilderness character, such as an impression from noticeable ecological impacts (natural quality) or the permanent closure of an area. Long-term physical effects to the wilderness character are 10 to 20 years.

Box 8 — Hypothetical Examples of Describing the Effects

Programmatic

The standards for stream bank trampling in this alternative (no more than 50 feet of human caused stream bank erosion to occur in any ¼ mile stretch of stream reach) would reduce impacts to the natural quality of wilderness character by providing a quantifiable measure that articulates when activities or uses should be prohibited, suspended, or limited.

There would be long term, moderate to major beneficial effects to the stream function and riparian vegetation (naturalness quality) by reducing increasing erosion along stream banks that is occurring as a result of poor trail alignment at stream crossings.

Project

There could be moderate, short term impacts to the naturalness quality of wilderness character when the stream crossing rehabilitation work is being conducted. The stream may at first experience a change from its channel, which through erosion, has deepened and widened and supported habitat for bank swallows. In the long term, this stream channel will return to more natural conditions and there will be a major beneficial effect to the naturalness quality.

Cumulative effects—Before addressing the cumulative effects for any project, different approaches need to be discussed with the district or forest planner. One of the first things to consider is if the analysis area is different for cumulative effects than for direct and indirect effects. It usually is different. For example, if there are adjacent or contiguous wilderness units, the cumulative effects analysis should include actions that have taken place on these adjacent wilderness lands. An example of a cumulative effects analysis is given in Box 9.

Box 9 — Hypothetical Example of a Cumulative Effects Analysis for a Project

This proposed action, to place a footbridge over the Clear Creek stream, when added to all past and present actions, will have minor cumulative effects on the undeveloped quality of wilderness character. There have not been any past actions in the vicinity of the footbridge, and no present or reasonable future actions are proposed within a 5-mile radius of this location. Wilderness-wide, there are 12 such footbridges, therefore this has an additive minor effect to the undeveloped quality, and is minor because these footbridges are handhewn logs and project a primitive quality.

Wilderness-wide, constructing and placing this footbridge, in addition to the other existing structures (including two drift fences and one other footbridge, neither in the same proximity), may have minor adverse effects to the undeveloped wilderness quality. When viewed collectively with the various structures from past uses, remnants of past recreational impacts, and current proposed allowable uses of commercial pack stock activities, very few areas may have an appearance of human occupation and improvements. This effect may be long-term and range from negligible to moderate intensity relative to a person's perceptions.

Often, interdisciplinary teams collectively catalogue past, present, and reasonably future actions. This catalogue is then considered by each specialist when drafting the cumulative effects analysis. Table 5 provides an example of how to catalogue potential actions to be considered in an analysis.

Table 5—Examples showing how the effects of different actions or events may affect different resources and how these effects may be catalogued.

Action or event	Summary of effects	Possible resources affected
Heidel Dam (past to 1928)	Inundation of habitat; change of vegetation; untrammeled, natural, and undeveloped qualities of wilderness character degraded	Botany, fish, watershed, soils, wilderness, wildlife
Fire suppression (past and present)	Vegetation composition altered; untrammeled, natural qualities of wilderness character degraded	Botany, wilderness
Recreational day use (present)	Opportunities for solitude quality of wilderness character degraded	Wilderness

When addressing cumulative effects, this catalogue becomes the reference from which the current proposal is measured. Temporally, the analysis considers which past actions may be relevant and when this action is added. The effects of this action, combined with the past effects, create additional effects.

In some cases, it may be impossible to accomplish the Purpose and Need of the project without some adverse effects to some component of wilderness character. This is especially true when the project falls under the Wilderness Act Section 4(d) special provisions categories, where proponent-driven projects may seemingly incongruous with wilderness objectives. In many cases, there may be an adverse effect to one wilderness character quality, such as untrammeled, but the net result is an overriding beneficial effect to some other quality, such as naturalness. This is a common occurrence, where Agency actions might "trammel" the wilderness resource in order to achieve the stated Purpose and Need to preserve or improve some other wilderness quality.

Decision Document

The main content of a decision document is a record of the decision, any mitigation and monitoring that needs to occur, and the decisionmaker's rationale for the decision (whether it is a Record of Decision for an Environmental Impact Statement, Decision Notice for an Environmental Assessment, or a Decision Memo for a Categorical Exclusion). It may be useful to discuss whether particular indicators of wilderness character are stable, declining, or improving under the decision. The chart comparing alternatives in *Appendix C* provides a mechanism for summarizing the effects of projects of programmatic analysis that could be a helpful tool in writing the rationale discussion. The net effect to wilderness character for all proponent driven projects should at a minimum be stable. For all projects, the effects to various aspects of wilderness character should be stable or improving. A hypothetical example addressing the trade-offs between two qualities of wilderness character is given in Box 10.

Box 10 — Hypothetical Example of the Trade-Offs Between Two Qualities of Wilderness Character

A project is proposed to control an invasive weed species that is beginning to establish on the edge of the No Name Wilderness. This is a new invasive species that has not been previously documented in the No Name Wilderness but has the potential to spread rapidly and seriously affect the naturally functioning ecosystem. The Agency has proposed to treat this infestation with herbicides—an action that would trammel the "earth and its community of life" inside wilderness. However, controlling the weed would have tangible positive effects on maintaining a naturally functioning ecosystem. This trade-off must be clearly documented and the aggregate positive benefit from the project described.

Monitoring Trend in Wilderness Character at the Local Level

Stewardship of individual wildernesses can be enhanced by making periodic assessments of the trend in wilderness character. Such assessments provide insight into progress in wilderness stewardship and suggest where improvement is most needed. This can help focus planning and management activities, increasing efficiency and effectiveness.

Previous work to develop a Forest Service (Landres and others 2005) and interagency framework for monitoring wilderness character provides a list of indicators that need to be evaluated to assess this trend. Although the interagency framework suggests several measures that can be useful in evaluating each indicator (*Appendix A*), locally derived measures would be more relevant and useful in making these evaluations.

Process to Monitor Trend in Wilderness Character

The overall process to monitor the trend in wilderness character is:
1. Review the list of indicators and make modifications as appropriate.
2. For each indicator, identify measures that can be used to assess trend in the indicator.
3. Assess trend for each measure.
4. Use these assessments of trend for each *measure* to decide if each indicator is improving, degrading, or stable.
5. Use these assessments of trend for each *indicator* to decide if each quality of wilderness character is improving, degrading, or stable.
6. Use these assessments of trend for each *quality* to decide whether overall wilderness character is improving, degrading, or stable.

Step 1: Review the list of indicators—The list of indicators from the interagency framework (*Appendix A*) provides a relatively complete listing of potentially relevant indicators. However, additional indicators can be added if local staffs feel there is something missing. It is also possible that some of these indicators are not applicable in all wildernesses. For example, status and trends in water quality are not applicable in a wilderness with no water bodies. The product of this step is a final list of applicable indicators.

Step 2: Identify measures for each indicator—Trends in each indicator will be assessed using one or more measures for each indicator. While there is no absolute minimum number of measures, we suggest that there should be at least two indicators for each quality (more if possible). Measures can be either quantitative or qualitative. For example, to assess trends in use of motorized equipment (one of the indicators of the undeveloped quality of wilderness character), some wildernesses may have counts of the number of times that motorized equipment has been used in the past year. If so, they can use these counts to assess whether the

USDA Forest Service Gen. Tech. Rep. RMRS-GTR-217WWW. 2008

21

number of incidents is increasing (degrading indicator), decreasing (improving indicator), or staying about the same (stable indicator). Other wildernesses will not have such counts and the best available source of information might be the judgments of one or several people regarding whether the number of incidents has increased, decreased, or stayed the same. Quantitative measures are preferable to qualitative measures because they make it easier to compare assessments made at different times and by different evaluators. However, it is common to lack quantitative data for all of the most important attributes of wilderness character. It is generally better to provide qualitative assessments of an important attribute than to ignore it.

To be most useful, local assessments of wilderness character should be based on measures that are locally relevant. The measures included in the national framework (*Appendix A*) were selected for their national applicability. They may not reflect the most important measures at the local level. Numerous lists of potentially useful monitoring parameters also exist. Enabling legislation, existing plans, other mandates, policies, and guiding documents can also be sources of ideas. Local expert opinion can be tapped—both to decide on measures and identify suitable protocols for quantifying measures. The products of this step are the measures used to assess each indicator and a protocol for how each measure is to be assessed.

Step 3: Assess trend for each measure—For each measure, decide whether conditions are improving, degrading, or staying the same. For quantitative measures, this analysis involves comparing quantities at two points in time and deciding if the difference is large enough to constitute significant improvement or degradation. For example, a 1 percent increase in visitation might be considered too close to zero to be considered an increase (measure is stable), while a 10 percent increase would be considered an increase (degradation of measure). Many qualitative assessments will be direct evaluations of trend. For example, experts might conclude that visitation is higher than in the past even though there are no counts to use as the basis for judgment.

There are no rules regarding how frequently measures or overall wilderness character should be assessed. Some measures are highly variable (such as air quality), while others are highly stable (such as number of dams). The frequency in taking measures should probably be dictated by the value of monitoring to make stewardship decisions rather than for the purpose of character monitoring. An assessment of trends in overall character every 5 years is probably adequate. The product of this step is a trend assessment of either "improving," "degrading," or "stable" for each measure.

Step 4: Assess trend for each indicator—The product of this step is an assessment of trend for each indicator. If there is only one measure for an indicator, the trend for the measure is the trend for the indicator. However, if there is more than one measure for each indicator, it will be necessary to aggregate several individual measure assessments into a single rating for the indicator. There are many ways to do this, each with pros and cons. There are three options for aggregation:

Equal Weighting—The approach taken by the Forest Service national framework and the interagency framework is to consider each measure to be of equal importance. For each measure, assign a rating of +1 if the trend improved, −1 if it degraded, and 0 if it was stable. Sum the ratings for all measures of each indicator. If this sum is a positive number, that indicator has improved and should be assigned a rating of +1. If the sum is negative, the indicator has degraded and should be assigned a rating of −1. If the sum is 0, the indicator is stable and should be assigned a rating of 0. In the Forest Service technical guide, a distinction is made between stable (when all measures are unchanged) and stable offsetting (when equal numbers of measures have improved and degraded).

Differential Weighting—An alternative approach is to consider the measures to vary in importance. More important measures can be assigned higher weightings than less important measures. The first step in this approach is to assign numerical weightings based on relative importance to each of the measures. For example, if the status of grizzly bears is considered to be twice as important to an assessment of the status of animal populations as the status of mountain goats, give the grizzly bear measure a weight of 2 and mountain goat measure a weight of 1. Then for each measure, assign a rating of +1 if it improved, −1 if it degraded, and 0 if it was stable. Take these +1, −1 and 0 ratings for each measure and multiply them by the weights of each measure. Sum these weighted ratings for all measures of each indicator. If this sum is a positive number, the indicator has improved and should be assigned a rating of +1. If the sum is negative, the indicator has degraded and should be assigned a rating of −1. If the sum is 0, the indicator is stable and should be assigned a rating of 0.

Holistic/Gestalt—A third approach is to simultaneously consider the trends in all the measures for each indicator and make a judgment regarding whether the overall trend for that quality is improving, degrading, or stable. The evaluator considers the number of measures that change, their relative importance, and the magnitude of change in making a final decision. This approach is more holistic than the more rigorously quantitative approaches described above, but different evaluators may arrive at different judgments. This suggests that it may be important to consider a group process when arriving at decisions. We also recommend writing a narrative on the rationale behind the final decision.

Step 5: Assess trend for each quality of wilderness character—Assessments from several indicators need to be aggregated into a single assessment for each of the four qualities of wilderness character. This process is similar to that of aggregating multiple measures into a single assessment for each indicator. It can be accomplished using equal weighting, differential weighting or a holistic/gestalt approach to aggregation. The product of this step is an assessment of trend for each of the four qualities.

Step 6: Assess overall trend in wilderness character—Assessments for the four qualities need to be aggregated into a single assessment for wilderness character. This process is similar to that of aggregating multiple measures into a single assessment of each indicator. Aggregation can occur on the basis of equal weighting, differential weighting, or by using a holistic/gestalt approach. The product of this step is an assessment of overall trend in wilderness character.

Monitoring and Information Management

Those interested in evaluating local monitoring data to evaluate trends in wilderness character in their wilderness will need to think through how these data will be collected, stored, and analyzed.

- *Data collection*—routine data collection protocols exist for monitoring some indicators but not all, and even existing protocols made need modifications to suit local conditions. If data are to be comparable over time, it is crucial that the procedures used to collect data be standardized and documented.
- *Data storage* — the storage of monitoring data in an electronic format is required for long-term protection and to facilitate analysis. Storage solutions can include locally developed databases, such as MS Access, as well as spreadsheets, like MS Excel. Corporate solutions, such as Infra and NRIS (Natural Resource Information System), are also available for some indicators and may offer practical advantages over locally developed applications, such as institutionalized support, integration with other data, and packaged analytical routines.
- *Analysis* — as with data collection, it is important to document the specific analytical procedures used to assess the condition of the indicator and trends over time. Familiarity with basic statistical procedures is a must. Again, these protocols need not be exhaustive but should be documented in enough detail so that someone unfamiliar with protocol could repeat the process in exactly the same steps.

Applying Wilderness Character to Managing Wilderness

In addition to planning and monitoring, wilderness character concepts can be useful in other aspects of wilderness stewardship. The following discussion is not exhaustive and readers are encouraged to think creatively of other appropriate applications of wilderness character.

Minimum Requirements Decision Guide

The Minimum Requirements Decision Guide (MRDG) was designed for use when making decisions regarding administrative decisions that involve "prohibited uses" listed in Section 4(c) of the Wilderness Act, such as the location of monitoring equipment installations inside wilderness or the use of motorized equipment, such as chainsaws. The two-step process helps managers: 1) determine if an administrative action in wilderness is necessary; and if it is, 2) determine the modifications that could be made to the proposed activity to minimize potential adverse impacts to wilderness values.

Though law or policy does not specifically require use of the MRDG, the Wilderness Act does require an assessment of what is necessary to meet "minimum requirements for the administration of the area for the purpose of the Act." The process can be used to assist with projects undergoing NEPA analysis or it can be used independent of the NEPA process.

In recent years, the MRDG was modified to incorporate the concept of wilderness character into the analysis. In Step 1, each potential administrative action needs to be evaluated based on its potential to affect wilderness character. Those responsible for completing the MRDG are asked to evaluate and document whether the proposed activity will preserve or impair each of the four qualities that comprise wilderness character ("Untrammeled," "Undeveloped," "Natural," and "Solitude or a primitive and unconfined type of recreation") and to document the reasoning for this assertion. A place is also provided to document effects to "other unique components that reflect the character of this wilderness." In Step 2, the effects of the each alternative for the proposed activity then need to be evaluated against each of the four qualities of wilderness character.

For example, if an MRDG were conducted to evaluate potential actions to remove aircraft wreckage from wilderness, the following qualities might be involved and the effects discussed:

- *Untrammeled* — options for addressing the wreckage would typically be quite small in scope and effects to the untrammeled quality would likely not be evaluated;
- *Natural* — aircraft materials are clearly foreign to wilderness ecosystems, and their presence and deterioration would have the potential to alter natural conditions, albeit on a very limited scale;

- *Undeveloped*—leaving the wreckage in wilderness would provide physical evidence of modern human presence; and
- *Solitude or a primitive and unconfined type of recreation* — viewing the wreckage would likely have a negative impact on one's sense of the primitive nature of wilderness.

Use of the MRDG doesn't guarantee that the management activities that are conducted in wilderness are "necessary to meet minimum requirements for the administration of the area for the purpose of the Act," or that wilderness character will be preserved. But it does provide a framework for a thoughtful and thorough discussion of the potential effects of an administrative action on wilderness character.

Those wanting to learn more about the Minimum Requirements Decision Guide process should go to http://www.wilderness.net/mrdg/ and on-line training on the Minimum Requirements Decision process is available at: http://www.carhart.wilderness.net/.

Fire Resource Advisor

The role of the Fire Resource Advisor (READ) working in wilderness is to ensure that the Line Officer's wilderness fire management objectives, resource concerns, and constraints on suppression activities are communicated, understood, and implemented by the Incident Management Team (IMT). The READ works for the line officer and directly with various members of the ICT as needed to help plan strategies and incorporate specific guidelines, practices, and alternative actions. The goal is to meet both wilderness and fire management objectives with minimal human caused disturbance. The READ is typically a local wilderness manager or ranger with knowledge of the area and the elements of the wilderness resource requiring special protection.

The concept of wilderness character can be a significant part of the READ's responsibilities. Of greatest concern are the effects of fire suppression activities and Burned Area Emergency Rehabilitation (BAER) or other post-fire restoration work. Each of the four qualities of wilderness character is used to help formulate guidelines and screen activities. Examples include:

- *Untrammeled*—minimizing and mitigating BAER or other restoration work such as erosion control or seeding;
- *Natural*—preventing the introduction of non-native invasive species during suppression activities, BAER, or restoration;
- *Undeveloped* — limiting the use of motorized equipment; and
- *Solitude or a primitive and unconfined type of recreation* — minimizing and restoring the impacts of suppression activities.

Information Needs Assessment

An Information Needs Assessment (INA) is defined as "a structured approach for determining data collection, storage, and analysis needs by first identifying and prioritizing local management requirements." INAs are conducted to ensure information is available, of sufficient quality and in the right format, to support key decisions related to wilderness stewardship while making most efficient use of limited resources.

Though an INA is not mandated, its use is encouraged to ensure limited program resources are directed toward efforts that yield the most useful information for wilderness stewardship purposes. Element 9 (Information Management) of the 10-Year Wilderness Stewardship Challenge (Wilderness Advisory Group 2008) recognizes this value by awarding two points to those units that have completed an INA for a particular wilderness.

The concept of wilderness character dovetails closely with the INA process. Those conducting an INA are instructed to identify the issues of concern for a particular wilderness. For this task, it may be useful to develop a "threats matrix" (Cole 1994) or some version of it. The threats matrix arrays the potential threats to wilderness resources, such as livestock grazing and recreation overuse, against the effects of those threats to the attributes of wilderness character. Effort should be made to ensure that the attributes included address all four qualities of wilderness character and not just biophysical resources. For example, the "number of agency actions that manipulate wilderness resources" might be tracked over time to evaluate trends in the untrammeled quality or the "number and type of recreation facilities" might be monitored for their effects on the solitude or primitive and unconfined type of recreation quality.

Work Planning

If a local monitoring approach were implemented to evaluate wilderness character in a particular wilderness, the resulting information on the trends in wilderness character and its component qualities and indicators would be useful in the identification and prioritization of potential work items necessary to preserve, or perhaps even improve, wilderness character. Trends in indicators demonstrating degrading conditions would be logical choices for investing the resources to stop or reverse these trends.

It is also important to communicate to those within the agency and the public the need to focus limited resources on stewardship actions that address degrading conditions related to wilderness character. Implementing specific on-the-ground actions that are aimed at improving wilderness character provides an excellent opportunity to explain the foundational concept of wilderness character and the framework that has been developed to identify its component parts.

References

American Heritage Dictionary, fourth ed. 2006. Houghton Mifflin Company, Boston, MA. 2112 p.

Aplet, G.H. 1999. On the nature of wildness: exploring what wilderness really protects. Denver University Law Review 76:347-367.

Belsky, A.J.; Blumenthal, D.M. 1997. Effects of livestock grazing on stand dynamics and soils in upland forests of the Interior West. Conservation Biology 11:315-327.

Borrie, W.T. 2000. Impacts of technology on the meaning of wilderness. In: Watson, A.E.; Aplet, G.H.; Hendee, J.C., comps. Personal, Societal, and Ecological Values of Wilderness: Sixth World Wilderness Congress Proceedings on Recreation, Management, and Allocation, Volume II; 1998 Oct 24-29, Bangalore, India. Proc. RMRS-P-14. Fort Collins, CO: U.S. Department of Agriculture, Forest Service, Rocky Mountain Research Station: 87-88.

Borrie, W.T.; Birzell, R.M. 2001. Approaches to measuring quality of the wilderness experience. In: Friedmund, W.A.; Cole, D.N., comps. Visitor Use Density and Wilderness Experience: Proc. RMRS-P-20. Fort Collins, CO: U.S. Department of Agriculture, Forest Service, Rocky Mountain Research Station: 29-38.

Cole, D.N. 1994. The wilderness threats matrix: a framework for assessing impacts. Res. Pap. INT-475. Ogden, UT: U.S. Department of Agriculture, Forest Service, Intermountain Research Station.

Cole, D.N.; Petersen, M.E.; Lucas, R.C. 1987. Managing wilderness recreation use: common problems and potential solutions. Gen. Tech. Rep. INT-230. Ogden, UT: U.S. Department of Agriculture, Forest Service, Intermountain Research Station.

Cowell, C.M.; Dyer, J.M. 2002. Vegetation development in a modified riparian environment: human imprints on an Allegheny River wilderness. Annals of the Association of American Geographers 92:189-202.

Dustin, D.L.; McAvory, L.H. 2000. Of what avail are forty freedoms: the significance of wilderness in the 21st century. International Journal of Wilderness 6(2):25-26.

Hall, T.E. 2001. Hikers' perspectives on solitude and wilderness. International Journal of Wilderness 7(2):20–24.

Hendee, J.C.; Dawson, C.P. 2002. Wilderness management: stewardship and protection of resources and values, third edition. Golden, CO: Fulcrum Publishing. 640 p.

Humphrey, H.H. 1957. Testimony June 19-20 for the U.S. Congress, Senate Committee on Interior and Insular Affairs, published hearings on S. 1176.

Keeling, P.M. 2007. Beyond the symbolic value of wilderness. International Journal of Wilderness 13(1):19-23.

Knapp, R.A.; Corn, P.S.; Schindler, D.E. 2001. The introduction of nonnative fish into wilderness lakes: good intentions, conflicting mandates, and unintended consequences. Ecosystems 4:275-278.

Landres, P. 2003. Database of Wilderness Laws and Their Management Language. Unpublished Excel database on file at: U.S. Department of Agriculture, Forest Service, Rocky Mountain Research Station, Aldo Leopold Wilderness Research Institute, Missoula, MT.

Landres, P.; Barns, C.; Dennis, J.G.; Devine, T.; Geissler, P.; McCasland, C.S.; Merigliano, L.; Seastrand, J.; Swain, R. 2008. Keeping it wild: an interagency strategy to monitor trends in wilderness character across the National Wilderness Preservation System. Gen. Tech. Rep. RMRS-GTR-212. Fort Collins, CO: U.S. Department of Agriculture, Forest Service, Rocky Mountain Research Station. 85 p.

Landres, P.; Boutcher, S.; Merigliano, L.; Barns, C.; Davis, D.; Hall, T.; Henry, S.; Hunter, B.; Janiga, P.; Laker, M.; McPherson, A.; Powell, D.S.; Rowan, M.; Sater, S. 2005. Monitoring selected conditions related to wilderness character: a national framework. Gen. Tech. Rep. RMRS-GTR-151. Fort Collins, CO: U.S. Department of Agriculture, Forest Service, Rocky Mountain Research Station. 38 p.

Landres P.; Marsh, S.; Merigliano, L.; Ritter, D.; Norman, A. 1998. Boundary effects on national forest wildernesses and other natural areas. In: Knight, R.L.; Landres. P.B., eds. Stewardship Across Boundaries. Washington, DC: Island Press: 117-139.

Landres, P.; Boutcher, S.; Hall, T.; Dean, L.; Mebane, A.; Blett, T.; Merigliano, L. [In press]. Technical guide for monitoring selected conditions related to wilderness character. Washington, DC: U.S. Department of Agriculture, Forest Service.

Leopold, A. 1921. The wilderness and its place in forest recreational policy. Journal of Forestry 19(7):718-721.

Leopold, A. 1949. A sand county almanac and sketches here and there. London, England: Oxford University Press. 240 p.

Louda, S.M.; Stiling, P. 2004. The double-edged sword of biological control in conservation and restoration. Conservation Biology 18:50-53.

Lucas, R.C. 1973. Wilderness: a management framework. Journal of Soil and Water Conservation 28:150-154.

Lucas, R.C. 1983. The role of regulations in recreation management. Western Wildlands 9(2):6-10.

Manning, R.E.; Lime, D.W. 2000. Defining and managing the quality of wilderness recreation experiences. In: McCool, S.F.; Cole, D.N.; Borrrie, W.T.; O'Loughlin, J., comps. Wilderness science in a time of change conference, Volume 4: Wilderness visitors, experiences, and visitor management. Proc. RMRS-P-15-VOL-4. Ogden, UT: U.S. Department of Agriculture, Forest Service, Rocky Mountain Research Station: 13-52.

McCloskey, M. 1999. Changing views of what the wilderness system is all about. Denver University Law Review 76:369-381.

McDonald, B.; Guldin, R.; Wetherhill, R. 1989. The spirit of wilderness: the use and opportunity of wilderness experience for personal growth. In: Freilich, H.R., comp. Wilderness benchmark 1988: proceedings of the national wilderness colloquium. Gen. Tech. Rep. SE-51. Asheville, NC: U.S. Department of Agriculture, Forest Service, Southeastern Forest Experiment Station: 193-207.

Roggenbuck, J.W. 2004. Managing for primitive recreation in wilderness. International Journal of Wilderness 10(3):21–24.

Rohlf, D.; Honnold, D.L. 1988. Managing the balance of nature: the legal framework of wilderness management. Ecology Law Quarterly 15:249-279.

Schreiber, R.K.; Newman, J.R. 1987. Air quality in wilderness: a state-of-knowledge review. In: Lucas, R.C., comp. National Wilderness Research Conference: Issues, State-of-Knowledge, Future Directions. Gen. Tech. Rep. INT-220. Ogden, UT: U.S. Department of Agriculture, Forest Service, Intermountain Research Station: 104-134.

Schroeder, H.W. 2007. Symbolism, experience, and the value of wilderness. International Journal of Wilderness 13(1):13-18.

Scott, D.W. 2002. "Untrammeled," "wilderness character," and the challenges of wilderness preservation. Wild Earth 11(3/4):72-79.

Sutter, P. 2004. Driven wild: how the fight against automobiles launched the modern wilderness movement. Seattle, WA: University of Washington Press. 343 p.

USDA Forest Service. 1972. Wilderness Policy Review, May 17, 1972. Worf, W.A.; Gorgensen, C.G.; Lucas, R.C., authors. Unpublished document on file at: U.S. Department of Agriculture, Forest Service, Rocky Mountain Research Station, Aldo Leopold Wilderness Research Institute, Missoula, MT. 56 p. plus appendices.

United States Congress. 1983. U.S. House Report 98-40 from the Committee on Interior and Insular Affairs, March 18, page 43.

Watson, A.E. 1995. Opportunities for solitude in the Boundary Waters Canoe Area Wilderness. Northern Journal of Applied Forestry 12(1):12-18.

Wilderness Advisory Group. 2008. U.S. Forest Service Chief's 10-Year Wilderness Stewardship Challenge Guidebook, [Online]. Available: http://www.wilderness.net/NWPS/documents/FS/guidebook.doc [2008, June 27].

Zahniser, H. 1956. The need for wilderness areas. The Living Wilderness 59(Winter to Spring):37-43.

Zahniser, H. 1963. Editorial: Guardians not gardeners. The Living Wilderness 83(Spring to Summer):2.

Appendix A—Indicators and Example Measures for the Four Qualities

The indicators and measures shown in table 6 are derived from the Forest Service national framework (Landres and others 2005), Technical Guide (Landres and others, in press), and Interagency Strategy for Monitoring Trends in Wilderness Character (Landres and others 2008). These indicators and measures represent a national perspective and some may not be relevant in a particular wilderness or for a particular planning decision or project. For most indicators, several example measures are given here and forest staff may choose among them or develop others that are more relevant to their local needs, conditions, and wilderness legislation.

Table A1 —Indicators and example measures for the four qualities of wilderness character.

Quality	Indicator	Example measures
Untrammeled – Wilderness is essentially unhindered and free from modern human control or manipulation	Actions authorized by the Federal land manager that manipulate the biophysical environment	Number of actions to manage plants, animals, pathogens, soil, water, or fire
		Percent of natural fire starts that received a suppression response
		Number of lakes and other water bodies stocked with fish
	Actions not authorized by the Federal land manager that manipulate the biophysical environment	Number of unauthorized actions by other Federal or State agencies, citizen groups, or individuals that manipulate plants, animals, pathogens, soil, water, or fire
Natural – Wilderness ecological systems are substantially free from the effects of modern civilization	Plant and animal species and communities	Abundance, distribution, or number of indigenous species that are listed as threatened and endangered, sensitive, or of concern
		Number of extirpated indigenous species
		Number of non-indigenous species
		Abundance, distribution, or number of invasive non-indigenous species
		Number of acres of authorized active grazing allotments and number of animal unit months (AUMs) of actual use inside wilderness
		Change in demography or composition of communities
	Physical resources	Visibility based on average deciview and sum of anthropogenic fine nitrate and sulfate
		Ozone air pollution based on concentration of N100 episodic and W126 chronic ozone exposure affecting sensitive plants
		Acid deposition based on concentration of sulfur and nitrogen in wet deposition
		Extent and magnitude of change in water quality
		Extent and magnitude of human-caused stream bank erosion
		Extent and magnitude of disturbance or loss of soil or soil crusts

Quality	Indicator	Example measures
	Biophysical processes	Departure from natural fire regimes averaged over the wilderness
		Extent and magnitude of global climate change based on change in timing of greening from MODIS satellite imagery, glacial retreat from photopoints, change in temperature and precipitation patterns from RAWS data, change in snow depth from SNOTEL data, coastal erosion or accretion from photopoints, or change in the distribution of select plant communities (for example, treeline) from photopoints
Undeveloped – Wilderness retains its primeval character and influence, and is essentially without permanent improvement or modern human occupation	Non-recreational structures, installations, and developments	Index of authorized physical development
		Index of unauthorized (user-created) physical development
	Inholdings	Area and existing or potential impact of inholdings
	Use of motor vehicles, motorized equipment, or mechanical transport	Type and amount of administrative and non-emergency use of motor vehicles, motorized equipment, or mechanical transport
		Type and amount of emergency use of motor vehicles, motorized equipment, or mechanical transport
		Type and amount of motor vehicle, motorized equipment, or mechanical transport use not authorized by the Federal land manager
	Loss of statutorily protected cultural resources	Number and severity of disturbances to cultural resources
Solitude or Primitive and Unconfined Recreation – Wilderness provides outstanding opportunities for solitude or primitive and unconfined recreation	Remoteness from sights and sounds of people inside the wilderness	Amount of visitor use
		Number of trail contacts
		Number and condition of campsites
		Area of wilderness affected by access or travel routes that are inside the wilderness
	Remoteness from occupied and modified areas outside the wilderness	Area of wilderness affected by access or travel routes that are adjacent to the wilderness
		Night sky visibility averaged over the wilderness
		Extent and magnitude of intrusions on the natural soundscape
	Facilities that decrease self-reliant recreation	Type and number of agency-provided recreation facilities
		Type and number of user-created recreation facilities
	Trail development level	Number of trail miles in developed condition classes
	Management restrictions on visitor behavior	Type and extent of management restrictions

Appendix B—Hypothetical Decision Memo Scenario

This appendix describes a scenario, from start to finish, in which a decision memo is used to approve a project in the hypothetical Passerine Wilderness.

Scenario

A wilderness ranger volunteer on patrol reported that numerous sections of trail to Solitaire Lake crossed wet meadows and streams and stream banks were eroding and trails deepening in the meadows. The District Ranger directs her staff to do some preliminary work on a proposal with the intent of going through a NEPA process and implementing this project when a funding source can be found.

Step 1: Identify Need for Change

Wilderness Manager Anne Dove convenes a meeting of district staff to develop the project. The biologist, wilderness manager, and trails technician are present. They draft a Need for Change document based on photos brought back from the trail person and the collective memory of the staff. The document describes the (1) Current Condition, (2) Existing Forest Plan Direction, (3) Need for Change, and (4) Recommended Actions.

Current condition—A non-system trail to Solitaire Lake is receiving moderate use. The trail is not designed and crosses through the edge of a moist meadow complex and two small streams. Use of the trail is increasing and the trail is incising due to vegetation loss and the compaction of meadow soils. Water is now coursing down the incised trail channel instead of dissipating across the meadow vegetation. At the first stream crossing, hikers and stock are eroding the stream bank causing sediment to reach the stream. The stream is approximately twice its natural width (by comparing it to upstream and downstream width to depth ratios). There is a lateral nick point upstream of this crossing that is vulnerable to becoming a headcut. A small population of pink-necked toads (*Bufo pinkus*), which is on the Region's sensitive species list, occurs about 100 yards upstream. It appears that users may cross in a variety of locations to avoid getting their feet wet.

Existing forest plan direction—The team outlines all pertinent directions from the Land and Resource Management Plan (LRMP), connecting standards and guidelines to the four wilderness qualities.

Untrammeled quality

- Ensure administrative actions are conducted in a manner that reduces the need for mechanical transportation or motorized equipment.

Natural quality

- Manage pink-necked toad breeding habitat to prevent disturbance.
- Minimize erosion of the physical structure and condition of stream banks and shorelines to sustain desired habitat diversity.
- Ensure meadow hydrologic function is not altered.

Undeveloped quality

- Assess the need for adding a system trail based on public need for access and for a minimum trail system that relies upon primitive trail conditions using structures for resource protection only.

Solitude and primitive or unconfined recreation quality

- Maintain low levels of hiker and stock use throughout the wilderness. Allow low to moderate use levels at popular lakes, including Duck Lake, Webb Lake, Bill Lake, Wing Lake, and Solitaire Lake.

Need for change—The natural quality of wilderness character is at risk of further degradation with continued erosion and headcutting that may have upstream effects on the pink-necked toad.

Solitaire Lake is popular, but because the trail leading to this lake is not a system trail, there are no funds to maintain it. The Forest Plan allows for this level of use at this lake and for a trail to be added to the system if warranted. This level of use would be within the standards and guidelines for maintaining "outstanding opportunities for solitude or a primitive and unconfined type of recreation."

Recommended actions—The following actions are recommended.

- Add a system trail to Solitaire Lake.
- Move 50 yards of the existing user-created trail from the meadow to the lodgepole pine forest.
- Rehabilitate the stream crossing.

Step 2: Develop a Proposal

The team takes their initial draft of the Need for Change documentation to the District Ranger who immediately recognizes the opportunity to enhance wilderness character with a trail project that includes a strong restoration component. She instructs her staff to more fully explore the proposal and develop objectives and actions. The team comes back to her with the following:

PROPOSAL

OBJECTIVES

Preserve the natural quality of wilderness character by:
- Reducing erosion and sedimentation of the stream caused by the existing trail
- Protecting instream beneficial uses
- Protecting and sustaining the diversity of aquatic habitats that are characteristic of the area.

Protect the primitive and unconfined recreation at Solitaire Lake by:
- Designing crossings so visitors stay on trails.

Actions

1. Add the trail to Solitaire Lake to the system.

2. Move a portion of the exiting trail. The trail crossing the wet meadow complex can be moved 50 feet to the south to avoid the meadow entirely. This will require about 100 yards of new trail construction.

3. Restore the vegetation in the former trail that crossed the wet meadow.

4. Design one main trail stream crossing. Protect the banks with rock and direct traffic to terrace steps. Terrace steps and retainers at banks of crossing and add water bars in the approach to divert water off the sensitive parts of the trail. Move rocks and improve the ford to narrow the footprint of these actions. Rehabilitate multi-trails and discourage redundant use by using native materials. Armor the stream bottom if necessary to protect it from disturbance.

Step 3: Prepare Proposal for Public Scoping

The District Ranger concurs with the proposal and directs the team to prepare a document that includes a Purpose and Need and the proposed action. The team edits the objectives and the Need for Change document into a "purpose and need" and further develops the proposal to provide the public a fuller description of the proposed project.

Purpose and need—The purpose of this project is to improve the trail to Solitaire Lake. Solitaire Lake is a popular lake and the trail is not on the system, yet receives moderate levels of public use. Direction in the Forest Plan allows for this level of use and for a trail to be added to the system if warranted. The natural quality of wilderness character is at risk of further degradation with continued erosion and headcutting that may have upstream effects on the pink-necked toad (*Bufo pinkus*), a sensitive species on the Region 7 sensitive species list. Trail work is needed to reduce erosion and sedimentation of stream courses caused by the existing trail system and to protect instream beneficial uses, maintain and enhance existing riparian plant communities, and enhance aquatic habitat.

The proposed project work will ensure that use is consistent with Forest Plan direction, including managing pink-necked toad breeding habitat to prevent disturbance and minimizing erosion to protect stream banks and sustain habitat diversity.

Proposed actions
1. Add Solitaire Lake trail to the Forest's trail system.
2. Re-route a portion of the existing trail. The trail crossing the wet meadow complex can be moved 50 feet to the south to avoid the meadow entirely. This will require about 100 yards of new trail construction in the adjacent lodgepole pine forest. The trail will be 12 inches wide and include two waterbars to limit the structures needed to maintain a primitive trail.
3. Restore the vegetation in the former trail that crosses through the wet meadow. This will include transplanting plugs of native vegetation (*Carex* spp.) from surrounding areas and mulching with native materials such as sand, grass, and sedge.
4. Design one main trail stream crossing. Protect the banks with rock and direct traffic to terrace steps. Terrace steps and retainers at the banks of the crossing. Add water bars in the approach to divert water off the sensitive parts of the trail. Move rocks and improve the ford to narrow the footprint of these actions. Rehabilitate multi-trails and discourage redundant use by using native materials. Armor the stream bottom if necessary to protect it from disturbance.

Step 4: Documentation of NEPA Decision

The District Ranger took the team's proposed action and scoped the proposal with interested parties. In the cover letter and news release, she indicated that she felt this would be a Decision Memo and it should be categorically excluded from further documentation based on FSH 19091.5 Category 31.2 *Construction and reconstruction of trails.* This category requires a project file and a Decision Memo. The District Ranger states in her scoping letter that preliminary findings suggest that there are no extraordinary circumstances. The project file and decision will document a determination that extraordinary circumstances related to the proposed action do not warrant further analysis and documentation in an EA or EIS.

All comments from the public supported the proposal. The Ranger, quite satisfied, asked the wilderness manager to draft a Decision Memo, prepare a document that evaluated wilderness as an extraordinary circumstance, and direct the district biologist to prepare a biological evaluation.

Step 5: A Finding That No Extraordinary Circumstances Exist

The wilderness manager knows that a good way to structure a finding on wilderness and the nonexistence of an extraordinary circumstance is to frame it with the four wilderness qualities of wilderness character. The finding is as follows:

The Wilderness Act (Public Law 88-577) requires that wilderness character be preserved. This section documents our finding that wilderness character will in fact be preserved with the proposed action and therefore, there is no extraordinary circumstance that precludes use of a categorical exclusion. As Forest Service Handbook 30.3 states, the mere presence of an extraordinary circumstance, in this case designated wilderness, does not preclude use of a CE. It is the degree of the potential effect of a proposed action on these resource conditions that determines whether extraordinary circumstances exist. Wilderness character combines biophysical and experiential qualities and is not explicitly defined in the Act. However, wilderness is defined in Section 2(c) and through this definition, the concept of wilderness character can be expressed as:

> A wilderness, in contrast with those areas where man and his
> own works dominate the landscape, is hereby recognized as an
> area where the earth and its community of life are untrammeled
> by man, where man himself is a visitor who does not remain.
> An area of wilderness is further defined to mean in the Act, an
> area of undeveloped Federal land retaining its primeval charac-
> ter and influence, without permanent improvements of human
> habitation, which is protected and managed so as to preserve

its natural conditions and which (1) generally appears to have been affected primarily by the forces of nature, with the imprint of man's work substantially unnoticeable; (2) has outstanding opportunities for solitude or a primitive and unconfined type of recreation; (3) has at least five thousand acres of land or is of sufficient size as to make practicable its preservation and use in an unimpaired condition; and (4) may also contain ecological, geological, or other features of scientific, educational, scenic, or historical value.

For the following reasons, the proposal for repair and restoration of the Solitaire Lake trail does not affect wilderness character and therefore no extraordinary circumstance exists.

1. The natural quality of wilderness will be enhanced by this project because erosion and sedimentation that is occurring in the stream system will be arrested. This will have beneficial effects to the stream system and without this project, the natural qualities are vulnerable to continued degradation.
2. The opportunities for a primitive and unconfined type of recreation will be achieved as a more sustainable trail system will enhance visitor experience and recreational enjoyment of the area.
3. The trail work itself will involve constructing two water bars that are minimal structures with low visibility. Because they are minimal in size and visibility, they will not have an effect on the developed quality of wilderness and are within the standards and guidelines for trail structures in wilderness.
4. There is nothing in the proposed action that will impede natural processes. In fact, natural processes will be restored with the revegetation of the meadow where trail incisement is capturing water flow when water should be more evenly dissipated throughout the meadow. The repaired stream crossing will maintain the correct width-to-depth ratio throughout the stream reach.

In the four qualities of wilderness character described above, the project, as proposed, enhances and does not degrade the qualities. It is for these reasons that this demonstrates a determination of no effect to wilderness, a potential extraordinary circumstance. The natural quality of wilderness character is at risk of further degradation with continued erosion and headcutting that may have upstream effects on the pink-necked toad.

Solitaire Lake is a popular lake. The trail is not on the system yet receives use and should be considered a system trail so that agency funds can be used to maintain it. Direction allows for this level of use and for a trail to be added to the system if warranted. It is within standards and guidelines for maintaining opportunities for solitude and a primitive or unconfined recreation.

UNITED STATES DEPARTMENT OF AGRICULTURE
FOREST SERVICE
Atlantic-Pacific Region 7

DECISION MEMO
Solitaire Lake Trail
Reconstruction and Restoration
Birders National Forest
Ducks and Geese Counties, Nevada

Decision

It is my decision to put the Solitaire Lake trail on the Birders National Forest trail system and to reconstruct and stabilize ½ mile of trail in the Passerine Wilderness. The intent of this project is to enhance the wilderness character of the area by improving the natural and primitive and unconfined recreation qualities of wilderness. This will be accomplished by (1) rerouting trail sections that are impacting riparian, streamside, and meadow environs and (2) revegetating and restoring rerouted sections.

This action is categorically excluded from documentation in an environmental impact statement or an environmental assessment because the project fits within category 31.2 (1) construction and reconstruction of trails. I have concluded that this decision is appropriately categorically excluded from documentation in an environmental impact statement or environmental assessment as it is corrective and moves the land toward the desired conditions, forest plan objectives, and wilderness objectives that are intended to preserve wilderness character.

My conclusion is based on information presented in this document and the entirety of the project record, which includes (1) a finding of no effect to wilderness, a potential extraordinary circumstance, (2) a Biological Evaluation that assesses the effect on federally listed threatened or endangered species or designated critical habitat, species proposed for Federal listing or proposed critical habitat, or Forest Service sensitive species, and (3) a Watershed field report. All documents support the conclusion that this project enhances the biological, physical, and social environment.

Extraordinary Circumstances Finding

1. <u>Threatened and Endangered Species or Their Critical Habitat</u> — In accordance with Section 7(c) of the Endangered Species Act, a list of the listed and proposed threatened or endangered species that may be present in the project area was requested from the U.S. Fish and Wildlife Service. The information indicated that there is no critical habitat for any plant species (Botany Biological Evaluation [BE]), aquatic species (Aquatics BE), or terrestrial wildlife species (Wildlife BE). It was determined that this decision will have "no effect" on listed species or their critical habitats.

  USDA Forest Service Gen. Tech. Rep. RMRS-GTR-217WWW. 2008

2. Floodplains, Wetlands, or Municipal Watersheds — The project area includes some small spring-associated moist meadows. There are no jurisdictional wetlands. The existing trail is not designed and crosses through at the edge of a moist meadow complex and two small streams. As trail use increases, the trail is incising due to vegetation loss, presence of soft meadow soils, and the high water table that is now causing water to reach the surface. At the first stream crossing, sedimentation is reaching the stream as hikers and stock erode the stream bank where they cross. The stream is approximately twice its natural width (by comparing it to upstream and downstream width to depth ratios). There is a lateral nick point upstream of this crossing that is vulnerable to becoming a headcut. This project will correct these concerns and have a beneficial effect to the small, moist meadow.

The district biologist determined that the project would have no effect on the beneficial water uses (Watershed Field Report, Project File, USFS 2005).

This decision should not result in significant floodplain, wetland, or municipal watershed impacts.

3. Congressionally Designated Areas – Wilderness — For the following reasons, the proposal for repair and restoration of the Solitaire Lake trail does not have an effect on wilderness character and therefore no extraordinary circumstance exists.
   - The natural quality of wilderness will be enhanced by this project because erosion and sedimentation that is occurring will be arrested. This will have beneficial effects to the stream system and without this project, the natural qualities are vulnerable to continued degradation.
   - The opportunities for a primitive and unconfined type of recreation will be achieved because a more sustainable trail system will enhance visitor experience and recreational enjoyment of the area.
   - The trail work itself will involve constructing two water bars that are minimal structures with low visibility. Because they are minimal in size and visibility, they will not affect the developed quality of wilderness and are within the standards and guidelines for trail structures in wilderness.
   - There is nothing in the proposed action that will impede natural processes. In fact, natural processes will be restored with the revegetation of the meadow where trail incisement is capturing water flow instead of dissipating it more evenly throughout the meadow. The repaired stream crossing will rectify a situation where natural processes are being impeded by restoring and maintaining the appropriate width to depth ratio throughout the stream reach.

   The project, as proposed, enhances and does not degrade the qualities of wilderness character described above. This demonstrates a determination of no effect to wilderness, a potential extraordinary circumstance.

4. Inventoried Roadless Areas — The project area does not include any Inventoried Roadless Areas.

5. <u>Research Natural Areas</u> — There are no Research Natural Areas within or near the project boundaries.

6. <u>Native American Religious or Cultural Sites, Archaeological Sites, or Historic Properties or Areas</u> — The Forest sent a copy of the proposed action to the local tribal council. No tribal concerns were identified.

   Heritage resource surveys were conducted during a 2004 field visit and an earlier (1995) project by a paraprofessional archaeologist. No heritage resources were identified within the area of potential effect. If heritage resources are encountered during project implementation, work will stop until Forest Heritage Resources staff can survey the site.

7. <u>Other Extraordinary Circumstances</u> — No other extraordinary circumstances related to the project were identified during internal and external scoping of the proposed action (Project Record, USFS 2005).

Project Description

The purpose of this project is to improve the trail to Solitaire Lake. Solitaire Lake is a popular lake and the trail is not on the system yet receives moderate levels of public use. Direction in the Forest Plan allows for this level of use and for a trail to be added to the system if warranted. The natural quality of wilderness character is at risk of further degradation with continued erosion and headcutting that may have upstream effects on the pink-necked toad (*Bufo pinkus*), a sensitive species on the Region 7 sensitive species list. Trail work is needed to reduce erosion and sedimentation entering stream courses from the existing trail system and to protect instream beneficial uses, maintain and enhance existing riparian plant communities, and enhance aquatic habitat.

The proposed project work will ensure that use is consistent with Forest Plan direction, including managing pink-necked toad breeding habitat to prevent disturbance and minimizing erosion to protect stream banks and sustain habitat diversity.

Proposed Action

1. Add Solitaire Lake trail to the Forest's trail system.

2. Re-route a portion of the existing trail. The trail crossing the wet meadow complex can be moved 50 feet to the south to avoid the meadow entirely. This will require about 100 yards of new trail construction in the adjacent lodgepole pine forest. The trail will be 12 inches wide and include two waterbars to limit the structures needed to maintain a primitive trail.

3. Restore the vegetation in the former trail that crosses through the wet meadow. This will include transplanting plugs of native vegetation (*Carex* spp.) from surrounding areas and mulching with native materials such as sand, grass and sedge.

4. Design one main trail stream crossing. Protect the banks with rock and direct traffic to terrace steps. Terrace steps and retainers at the banks of the crossing. Add water bars in the approach to divert water off the sensitive parts of the trail. Move rocks and improve the ford to narrow the footprint of these actions. Rehabilitate multi-trails and discourage redundant use by using native materials. Armor the stream bottom if necessary to protect it from disturbance.

Public Involvement

The project was listed in the Birders National Forest Quarterly Schedule of Proposed Actions (SOPA) starting in September 2005. To initiate scoping for the project, a letter dated September 1, 2005 was sent to individuals and organizations who have expressed interest in trail projects on the Birders National Forest, including local Native American tribes and county, state, and federal agencies with jurisdiction in the area. A press release announcing the project was sent to the Daily News on September 2, 2005 and the proposed project was published on September 6, 2005.

Five letters were received. All comments were supportive of the project and there were no issues or disagreements with the proposal.

Findings Required by Other Laws

My decision will comply with all applicable laws and regulations. I have summarized the pertinent ones below.

1. Forest Plan Consistency (National Forest Management Act) — This action is consistent with the direction in the Birders National Forest Land and Resource Management Plan (LRMP) and its amendments. The project falls within the Passerine Wilderness Management Area and it is my determination that the project moves us toward desired wilderness character qualities, specifically for naturalness and opportunities for solitude and primitive or unconfined type of recreation.

2. National Environmental Policy Act — This Act requires public involvement and consideration of potential environmental effects. The entirety of documentation for this decision supports compliance with this Act.

3. Other Laws and Regulations — This project complies with all federal, state, and local laws and Executive Orders including the National Forest Management Act, Endangered Species Act (Project BE's), Clean Water Act (watershed report, Project File; CA Best Management Practices, see also II.B2 above), National Historic Preservation Act (see II.B6 above), Archaeological Resources Protection Act (see II.B6 above), Native American Graves Protection and Repatriation Act (see II.B6 above), Wild and Scenic Rivers Act (see II.B3 above), Executive Order 11990 (Wetlands) (see II.B2 above), Executive Order 11988 (Floodplains) (see II.B2 above), and Executive Order 12898 (Environmental Justice) (see III. Public Involvement, above; Project Record).

Implementation Date

It is my intention to implement this decision in the summer of 2007. Project implementation is dependent on funding. The district staff will be pursuing a variety of funding sources and partnerships to accomplish this work once a decision is made. The earliest date of implementation would be June 2007; the latest date would be October 2012.

Administrative Review or Appeal Opportunities

In accordance with the October 19, 2005 order issued by the U. S. District Court for the Eastern District of California in Case No. CIV F-03-6386JKS, this decision is subject to administrative appeal pursuant to 36 CFR Part 215. In accordance with the April 24, 2006 order issued by the U. S. District Court for the Division of the District of Nevada in Case No. CV 03-119-M-DWM, only those individuals and organizations who provided comments or otherwise expressed interest in the proposed action by the close of the comment period are eligible to appeal (36 CFR 215.11(a), 2002 version). Appeals must be filed within 45 days from the publication date of the legal notice in the Weekly Guardian Notices of appeal must meet the specific content requirements of 36 CFR 215.14. An appeal, including attachments, must be filed (regular mail, fax, e-mail, hand-delivery, express delivery, or messenger service) with the appropriate Appeal Deciding Officer (36 CFR 215.8) within 45 days following the publication date of the legal notice. The publication date of the legal notice is the exclusive means for calculating the time period to file an appeal (36 CFR 215.15 (a)). Those wishing to appeal should not rely upon dates or timeframe information provided by any other source.

Appeals must be submitted to Forest Supervisor, Birders National Forest, 351 Hawk Lane, Suite 200, Raptor, Nevada 93514, telephone 123.555.9999. Appeals may be submitted by FAX (123.555.2222) or by hand-delivery to the Forest Supervisor's Office, at the address shown above, during normal business hours of Monday through Friday 8:00am to 4:30pm. Electronic appeals, in acceptable plain text (.txt), rich text (.rtf), or Word (.doc) formats, may be submitted by email to appeals-pacificatlantic-birders@fs.fed.us with Subject: **Solitaire Lake**

Appendix C—Summarizing Effects Using Wilderness Character Indicators

An effects analysis quantifies potential impacts on the wilderness character of the area from each alternative of the proposed action. This effects analysis should assess impacts to all the indicators that are relevant to the proposed action and alternatives, ideally in a way that characterizes differences among alternatives and uses at least one indicator from each of the four qualities of wilderness character. Once the effects analysis is completed, the results would be summarized for each indicator, such as "Alternative 1: seven actions that manipulate the biophysical environment over a period of 1 year." This summary shows how each of the alternatives affects the four qualities of wilderness character. Table C1 offers a hypothetical example showing how this summary of effects would be organized by the four qualities of wilderness character.

Table C1—Summary of the effects of each alternative on the indicators of wilderness character. This table does not show the full set of possible indicators. Instead, it shows a subset of indicators determined by local staff to be affected by the different alternatives of the proposed action. These effects are organized by the four qualities of wilderness character.

Proposed action					
Quality	Component	Indicator	Summary of effects		
			Alt. 1	Alt. 2	Alt. 3
Untrammeled					
Wilderness is essentially unhindered and free from modern human control or manipulation	Authorized actions that control or manipulate the "earth and its community of life"	Actions authorized by the Federal land manager that manipulate the biophysical environment			
	Unauthorized actions that control or manipulate the "earth and its community of life"	Actions not authorized by the Federal land manager that manipulate the biophysical environment			

Proposed action					
Quality	Component	Indicator	Summary of effects		
			Alt. 1	Alt. 2	Alt. 3
Natural					
Wilderness eco-logical systems are substantially free from the effects of modern civilization	Terrestrial, aquatic, and atmospheric natural species and physical resources	Indigenous plant and animal species that are listed or of concern			
		Non-indigenous invasive plant and animal species			
		Water quality			
		Soil disturbance or erosion			
	Terrestrial, aquatic, and atmospheric biophysical processes	Departure from natural fire regimes			
Undeveloped					
Wilderness retains its primeval character and influence, and is essentially without permanent improvement or modern human occupation	Development	Non-recreational structures and Improvements			
	Mechanization	Motorized equipment use			
		Mechanical transport use			
	Loss of statutorily protected cultural resources	Disturbance to cultural sites			
Solitude or Primitive and Unconfined Recreation					
Wilderness provides outstanding opportunities for solitude or primitive and unconfined recreation	Outstanding opportunities for solitude	Remoteness from sights and sounds of people inside the wilderness			
	Outstanding opportunities for primitive and unconfined recreation	Management restrictions on visitor behavior			

In the absence of quantifiable effects, categories such as "moderate negative effect" could be used to summarize the effects of the alternatives on the indicators of wilderness character. Table C2 offers one scheme for such categories. Categories are not as explicit or accurate as quantifiable effects. But if the categories are defined in terms of intensity and duration of the effect (and geographical scope if appropriate), the reader or reviewer will at least have a better understanding of what these categories mean and the relative impacts from each of the different alternatives.

Table C2—Suggested categories for summarizing the effects of proposed alternatives on indicators of wilderness character.

Code	Category	Description
N/A	not applicable	This indicator is not applicable to this wilderness
***	significant negative effect	Effects are long lasting and have the potential to significantly degrade the wilderness character of the area
**	moderate negative effect	Effects are of moderate to long-term duration and have potential to appreciably degrade the wilderness character of the area
*	slight negative effect	Effects are of short-term duration and the effect on wilderness character is deemed negative though minor in intensity
0	no discernable effect	The effects of the proposed action on this indicator are negligible in intensity and duration
+	slight positive effect	Effects are of short-term duration and the effect on wilderness character is deemed positive though minor in intensity
++	moderate positive effect	Effects are of moderate to long-term duration and have potential to appreciably improve the wilderness character of the area
+++	significant positive effect	Effects are long lasting and have the potential to significantly improve the wilderness character of the area

The Rocky Mountain Research Station develops scientific information and technology to improve management, protection, and use of the forests and rangelands. Research is designed to meet the needs of the National Forest managers, Federal and State agencies, public and private organizations, academic institutions, industry, and individuals. Studies accelerate solutions to problems involving ecosystems, range, forests, water, recreation, fire, resource inventory, land reclamation, community sustainability, forest engineering technology, multiple use economics, wildlife and fish habitat, and forest insects and diseases. Studies are conducted cooperatively, and applications may be found worldwide.

Station Headquarters
Rocky Mountain Research Station
240 W. Prospect Road
Fort Collins, CO 80526
(970) 498-1100

Research Locations

Flagstaff, Arizona	Reno, Nevada
Fort Collins, Colorado	Albuquerque, New Mexico
Boise, Idaho	Rapid City, South Dakota
Moscow, Idaho	Logan, Utah
Bozeman, Montana	Ogden, Utah
Missoula, Montana	Provo, Utah